FINDING DINOSAURS

TRICERATOPS

by Samantha S. Bell

FOCUS READERS

FOCUS READERS

WWW.FOCUSREADERS.COM

Focus Readers is distributed by North Star Editions:
sales@northstareditions.com | 888-417-0195

Produced for Focus Readers by Red Line Editorial.

Content Consultant: Dr. David B. Weishampel, Professor Emeritus, Center for Functional Anatomy and Evolution, Johns Hopkins University School of Medicine

Photographs ©: Herschel Hoffmeyer/Shutterstock Images, cover, 1; Science Source, 4–5; *Popular Science Monthly* Volume 67/MBLWHOI Library, 7; Pavel L. Photo and Video/ Shutterstock Images, 8; Lyudmila Voronova/Shutterstock Images, 10–11; WaterFrame_sta/ Alamy, 13; Anton Ivanov/Shutterstock Images, 15; Millard H. Sharp/Science Source, 16; Kostyantyn Ivanyshen/Shutterstock Images, 18–19; Red Line Editorial, 20; Computer Earth/Shutterstock Images, 23; The Natural History Museum/Alamy, 25; Hayk Shalunts/ Shutterstock Images, 26–27; Rick Rudnicki/Alamy, 29

ISBN
978-1-63517-507-3 (hardcover)
978-1-63517-579-0 (paperback)
978-1-63517-723-7 (ebook pdf)
978-1-63517-651-3 (hosted ebook)

Library of Congress Control Number: 2017948119

Printed in the United States of America
Mankato, MN
November, 2017

ABOUT THE AUTHOR

Samantha S. Bell is an author and illustrator from South Carolina who has written more than 60 nonfiction books for children. She loves learning about dinosaurs and how they may have lived. She also enjoys being outside and digging in the dirt, but she has yet to discover any dinosaur bones.

TABLE OF CONTENTS

CHAPTER 1

BONE WARS

In the late 1800s, stories of dinosaur bones in Utah, Colorado, and Wyoming brought scientists from all over the United States. They all hoped to find and name the biggest and most exotic dinosaurs. Some scientists competed to find and name the most fossils. Their **rivalry** was known as the Bone Wars.

Othniel Charles Marsh (top) led many expeditions during the Bone Wars.

During the Bone Wars, scientists sometimes named fossils that other people found. They also made mistakes. For example, George Cannon found two large horns and a piece of a skull in 1887. He sent the bones to a famous scientist named Othniel Charles Marsh. Marsh wanted to find more of the skull. He sent workers to Denver, Colorado, where Cannon had found the bones. But they only found a few more horn fragments.

Marsh examined the bones Cannon had sent him. Marsh thought they belonged to a huge bison. But Cannon was not sure. It seemed odd for bison bones to be in the same rock as dinosaur bones.

John Bell Hatcher discovered more than 30 dinosaur skulls between 1889 and 1892.

In 1888, John Bell Hatcher was digging for dinosaur bones in Wyoming. He uncovered a dinosaur skull. The skull had long, pointed horns. Hatcher sent bones from the skull to Marsh. But Marsh did not realize that the bones came from the same kind of dinosaur as Cannon's find.

A Triceratops has been on display in the National Museum of Natural History in Washington, DC, since 1905.

Because the skull's horns did not match any known species, it appeared that Hatcher had discovered a new dinosaur.

Over the next three years, Hatcher found more skulls and bones from the same kind of dinosaur. He sent them all to Marsh. Marsh named the new dinosaur *Triceratops*, which means "three-horned face."

In 1904, scientists used the bones that Hatcher collected to put a Triceratops skeleton together. The skeleton was named Hatcher in his honor. Years later, other scientists decided that Cannon's original find was a Triceratops, too.

ASSEMBLING A SKELETON

Because no complete Triceratops skeleton was found, scientists combined bones from several specimens. They even used feet from a duck-billed dinosaur. In 1999, scientists worked to make the Hatcher skeleton more accurate. They used computers to figure out the size and shape of the missing bones.

CHAPTER 2

A THREE-HORNED GIANT

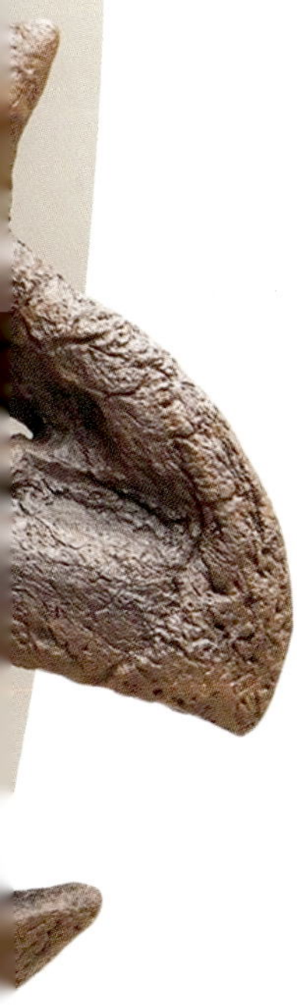

Scientists divide dinosaurs with horns into two groups. Dinosaurs with two large horns above their eyes and one small nose horn are called chasmosaurines. Triceratops is the most famous chasmosaurine. Dinosaurs with two small horns above their eyes and one large nose horn are called centrosaurines.

Scientists have uncovered more than 100 Triceratops skulls.

Centrosaurus and Styracosaurus are part of this second group.

Triceratops could be 10 feet (3.0 m) tall. It was approximately 30 feet (9.1 m) long. The dinosaur's skull made up almost one-third of the length of its body. The largest Triceratops skull ever found was 9.2 feet (2.8 m) long. This skull was 5.2 feet (1.6 m) high and 4.6 feet (1.4 m) wide.

Paleontologists studied the skulls of Triceratops specimens of different ages. They noticed that the two horns above the dinosaur's eyes changed as the animal grew. When a Triceratops was young, its horns curved backward. But as the dinosaur got older, its horns straightened

A paleontologist cleans a fossilized Triceratops horn.

out. Finally, the horns curved forward as the animal became an adult.

The horns of a baby Triceratops were approximately 1 inch (2.5 cm) long.

An adult's horns could be 3 feet (0.9 m) long. Triceratops may have used its horns to defend itself against **predators**. But scientists do not know for sure.

In 2015, scientists began studying a new species of three-horned dinosaur. Known as Regaliceratops, this dinosaur had two small horns above its eyes and one large nose horn. Regaliceratops could not have used its small horns for defense. They were just for display. Triceratops is closely related to Regaliceratops. For this reason, some scientists think Triceratops's horns were for display, too.

Triceratops had a bony frill behind its horns. The frill was approximately 6 feet

This replica shows what Triceratops may have looked like.

(1.8 m) wide. Scientists are still trying to determine what this frill was for. It might have helped Triceratops recognize one another. Or Triceratops might have used the frill to **regulate** its body temperature.

Fossilized remains help scientists study Triceratops skin.

If a Triceratops became too hot, extra body heat would be sent out from its frill into the air. The dinosaur would cool down as a result.

Males might have used their frills to attract mates, similar to how peacocks use their tails. Males might have fought to impress females, too. Some Triceratops frills have **puncture** marks. These marks probably came from other Triceratops's horns. The dinosaurs might have clashed to defend their territory or to show **dominance.**

SCALY SKIN

In 2002, a Triceratops specimen was found on a ranch in Wyoming. It included the largest piece of intact skin ever discovered. The skin had a pattern of scales shaped like hexagons. These scales were rougher and larger than the scales of many other dinosaurs. Some of the scales stuck out. Scientists are still not sure what these scales were for.

CHAPTER 3

LIVING ALONE

Triceratops lived during the Late Cretaceous period, approximately 66 million years ago. Fossils show that these horned dinosaurs lived in Canada and the western United States. Other dinosaurs such as Tyrannosaurus also lived in this area. Triceratops shared its habitat with mammals, birds, and insects as well.

Triceratops roamed across North America millions of years ago.

Most Triceratops fossils are of single individuals. For this reason, scientists think the dinosaur spent most of its life alone. However, several Triceratops fossils were found together in Wyoming in 2013. The fossils included one juvenile and three adults. This discovery showed that Triceratops may not have lived alone all the time.

GEOLOGICAL TIMELINE

Dinosaurs lived during the Mesozoic era. This era is divided into three different periods.

A full-grown Triceratops could weigh 4 to 6 tons (3.6 to 5.4 metric tons). The dinosaur carried its huge body on stout legs. Its front feet had five toes each. Its back feet had four toes. The dinosaur's back legs were longer than its front legs. Some scientists think Triceratops stood upright like an elephant. Others think its legs went outward, like those of a lizard. Scientists try to find the answer by studying the dinosaur's elbow bones.

Triceratops had a beak-like mouth. This mouth shape would have been better for plucking leaves than for biting prey. So, scientists believe Triceratops was an **herbivore**.

Triceratops probably ate a variety of plants, including ferns and palms. Triceratops's teeth helped it grind the plants it ate. The dinosaur had hundreds of teeth. They were arranged in columns in the cheek area. New teeth would push out the old, worn-down ones. The teeth were constantly being replaced. There may have been as many as four replacement teeth under every tooth. A Triceratops could have up to 800 teeth in its mouth.

Triceratops would have faced predators such as Tyrannosaurus. Some fossils of Triceratops frills show bite marks from Tyrannosaurus teeth. But Triceratops

Triceratops may have used its horns to defend itself against predators.

might have fought back. In 1997, scientists found a Triceratops fossil with one of the horns bitten off. Bite marks on the horn matched patterns left by Tyrannosaurus teeth. But the fossil also showed that the horn had healed. Scientists concluded that the Triceratops had survived the attack.

TRICERATOPS TEETH

In 2015, a team of scientists cut through several Triceratops teeth to examine their structure. They found that each tooth had five layers of **tissue**. Each layer had a different hardness. Teeth with more layers are better at breaking up food. Most reptile teeth have only two layers of tissue. The additional layers would have allowed Triceratops to eat more kinds of plants.

As the dinosaur ate, the tissues in its teeth would have worn down. To find out how the different tissues worked, scientists scratched the Triceratops teeth. The scratching imitated the way plants would have moved across the dinosaur's teeth. The scientists found that as the tissues wore down, the shape of the teeth changed. The scratching created many sharp surfaces. These

The shape of Triceratops's teeth allowed it to chew its food like a mammal does.

surfaces helped the teeth slice through thick plants. Engineers created a model to show how the tissues worked.

CHAPTER 4

STILL SEARCHING

Triceratops disappeared at the end of the Cretaceous period. Many other dinosaurs died out at this time. No one is sure why. But many scientists believe a huge meteor struck Earth. This meteor would have killed most living things.

Triceratops fossils have been found in Alberta and Saskatchewan in Canada.

Many Triceratops skeletons in museums are made up of fossils from several different dinosaurs.

They have also been found in Wyoming, Montana, North Dakota, South Dakota, and Colorado in the United States.

However, even the most complete Triceratops specimen had less than 80 percent of its skeleton. So, scientists sometimes use parts of other dinosaurs

RELATED DINOSAURS

Scientists also study fossils of similar dinosaurs. In 2015, two new species of horned dinosaurs were uncovered in Canada. Known as Regaliceratops and Wendiceratops, these dinosaurs are closely related to Triceratops. Studying these fossils could help scientists learn more about Triceratops as well.

The name *Regaliceratops* means "royal horned face."

to build the Triceratops skeletons that are displayed in museums.

Fossils of young Triceratops are also rare. But fossils such as the group found in Wyoming can help scientists learn how the dinosaur grew. Studying existing fossils and finding new specimens will help scientists continue to learn more.

FOCUS ON TRICERATOPS

Write your answers on a separate piece of paper.

1. Write a sentence summarizing how Triceratops may have used its horns.
2. Do you think scientists should use bones from different dinosaurs to build Triceratops skeletons? Why or why not?
3. Who gave Triceratops its name?
   - **A.** John Bell Hatcher
   - **B.** George Cannon
   - **C.** Othniel Charles Marsh
4. What might have happened if Triceratops had not had replacement teeth?
   - **A.** It would not have been able to chew its food as well.
   - **B.** It would not have been able to pluck leaves from plants.
   - **C.** It would not have been an herbivore.

Answer key on page 32.

GLOSSARY

dominance
Having more power or getting first choice of resources such as food, water, or mates.

herbivore
An animal that eats only plants.

paleontologists
Scientists who study the ancient past and the fossil remains of ancient living things.

predators
Animals that hunt other animals for food.

puncture
To make a hole in something.

regulate
To control the amount of something.

rivalry
When two people or groups compete against each other.

specimens
The remains of individual animals that show what the entire group of animals was like.

tissue
A group of similar cells in a plant or animal that have a certain job or function.

TO LEARN MORE

BOOKS

Gray, Susan H. *Triceratops*. Mankato, MN: The Child's World, 2015.

Hoena, Blake. *Everything Dinosaurs*. Washington, DC: National Geographic, 2014.

Thimmesh, Catherine. *Scaly Spotted Feathered Frilled: How Do We Know What Dinosaurs Really Looked Like?* Boston: Houghton Mifflin Harcourt, 2013.

NOTE TO EDUCATORS

Visit **www.focusreaders.com** to find lesson plans, activities, links, and other resources related to this title.

INDEX

Answer Key: 1. Answers will vary; **2.** Answers will vary; **3.** C; **4.** A